TERMS FOR DESCRIBING
ADVANCED NUCLEAR
POWER PLANTS

The following States are Members of the International Atomic Energy Agency:

AFGHANISTAN	GAMBIA	OMAN
ALBANIA	GEORGIA	PAKISTAN
ALGERIA	GERMANY	PALAU
ANGOLA	GHANA	PANAMA
ANTIGUA AND BARBUDA	GREECE	PAPUA NEW GUINEA
ARGENTINA	GRENADA	PARAGUAY
ARMENIA	GUATEMALA	PERU
AUSTRALIA	GUYANA	PHILIPPINES
AUSTRIA	HAITI	POLAND
AZERBAIJAN	HOLY SEE	PORTUGAL
BAHAMAS	HONDURAS	QATAR
BAHRAIN	HUNGARY	REPUBLIC OF MOLDOVA
BANGLADESH	ICELAND	ROMANIA
BARBADOS	INDIA	RUSSIAN FEDERATION
BELARUS	INDONESIA	RWANDA
BELGIUM	IRAN, ISLAMIC REPUBLIC OF	SAINT KITTS AND NEVIS
BELIZE	IRAQ	SAINT LUCIA
BENIN	IRELAND	SAINT VINCENT AND
BOLIVIA, PLURINATIONAL	ISRAEL	THE GRENADINES
STATE OF	ITALY	SAMOA
BOSNIA AND HERZEGOVINA	JAMAICA	SAN MARINO
BOTSWANA	JAPAN	SAUDI ARABIA
BRAZIL	JORDAN	SENEGAL
BRUNEI DARUSSALAM	KAZAKHSTAN	SERBIA
BULGARIA	KENYA	SEYCHELLES
BURKINA FASO	KOREA, REPUBLIC OF	SIERRA LEONE
BURUNDI	KUWAIT	SINGAPORE
CABO VERDE	KYRGYZSTAN	SLOVAKIA
CAMBODIA	LAO PEOPLE'S DEMOCRATIC	SLOVENIA
CAMEROON	REPUBLIC	SOUTH AFRICA
CANADA	LATVIA	SPAIN
CENTRAL AFRICAN	LEBANON	SRI LANKA
REPUBLIC	LESOTHO	SUDAN
CHAD	LIBERIA	SWEDEN
CHILE	LIBYA	SWITZERLAND
CHINA	LIECHTENSTEIN	SYRIAN ARAB REPUBLIC
COLOMBIA	LITHUANIA	TAJIKISTAN
COMOROS	LUXEMBOURG	THAILAND
CONGO	MADAGASCAR	TOGO
COSTA RICA	MALAWI	TONGA
CÔTE D'IVOIRE	MALAYSIA	TRINIDAD AND TOBAGO
CROATIA	MALI	TUNISIA
CUBA	MALTA	TÜRKİYE
CYPRUS	MARSHALL ISLANDS	TURKMENISTAN
CZECH REPUBLIC	MAURITANIA	UGANDA
DEMOCRATIC REPUBLIC	MAURITIUS	UKRAINE
OF THE CONGO	MEXICO	UNITED ARAB EMIRATES
DENMARK	MONACO	UNITED KINGDOM OF
DJIBOUTI	MONGOLIA	GREAT BRITAIN AND
DOMINICA	MONTENEGRO	NORTHERN IRELAND
DOMINICAN REPUBLIC	MOROCCO	UNITED REPUBLIC
ECUADOR	MOZAMBIQUE	OF TANZANIA
EGYPT	MYANMAR	UNITED STATES OF AMERICA
EL SALVADOR	NAMIBIA	URUGUAY
ERITREA	NEPAL	UZBEKISTAN
ESTONIA	NETHERLANDS	VANUATU
ESWATINI	NEW ZEALAND	VENEZUELA, BOLIVARIAN
ETHIOPIA	NICARAGUA	REPUBLIC OF
FIJI	NIGER	VIET NAM
FINLAND	NIGERIA	YEMEN
FRANCE	NORTH MACEDONIA	ZAMBIA
GABON	NORWAY	ZIMBABWE

The Agency's Statute was approved on 23 October 1956 by the Conference on the Statute of the IAEA held at United Nations Headquarters, New York; it entered into force on 29 July 1957. The Headquarters of the Agency are situated in Vienna. Its principal objective is "to accelerate and enlarge the contribution of atomic energy to peace, health and prosperity throughout the world".

IAEA NUCLEAR ENERGY SERIES No. NR-T-1.19

TERMS FOR DESCRIBING ADVANCED NUCLEAR POWER PLANTS

INTERNATIONAL ATOMIC ENERGY AGENCY
VIENNA, 2023

COPYRIGHT NOTICE

All IAEA scientific and technical publications are protected by the terms of the Universal Copyright Convention as adopted in 1952 (Berne) and as revised in 1972 (Paris). The copyright has since been extended by the World Intellectual Property Organization (Geneva) to include electronic and virtual intellectual property. Permission to use whole or parts of texts contained in IAEA publications in printed or electronic form must be obtained and is usually subject to royalty agreements. Proposals for non-commercial reproductions and translations are welcomed and considered on a case-by-case basis. Enquiries should be addressed to the IAEA Publishing Section at:

Marketing and Sales Unit, Publishing Section
International Atomic Energy Agency
Vienna International Centre
PO Box 100
1400 Vienna, Austria
fax: +43 1 26007 22529
tel.: +43 1 2600 22417
email: sales.publications@iaea.org
www.iaea.org/publications

Printed by the IAEA in Austria
October 2023
STI/PUB/2071

IAEA Library Cataloguing in Publication Data

Names: International Atomic Energy Agency.
Title: Terms for describing advance nuclear power plants / International Atomic
    Energy Agency.
Description: Vienna : International Atomic Energy Agency, 2023. | Series: IAEA
    nuclear energy series, ISSN 1995-7807 ; no. NR-T-1.19 | Includes bibliographical
    references.
Identifiers: IAEAL 23-01627 | ISBN 978-92-0-145923-7 (paperback : alk. paper) |
    ISBN 978-92-0-146023-3 (pdf) | ISBN 978-92-0-146123-0 (epub)
Subjects: LCSH: Nuclear reactors. | Nuclear power plants — Terminology. | Nuclear
    power plants — Design and construction.
Classification: UDC 621.311.25 | STI/PUB/2071

FOREWORD

The IAEA's statutory role is to "seek to accelerate and enlarge the contribution of atomic energy to peace, health and prosperity throughout the world". Among other functions, the IAEA is authorized to "foster the exchange of scientific and technical information on peaceful uses of atomic energy". One way this is achieved is through a range of technical publications including the IAEA Nuclear Energy Series.

The IAEA Nuclear Energy Series comprises publications designed to further the use of nuclear technologies in support of sustainable development, to advance nuclear science and technology, catalyse innovation and build capacity to support the existing and expanded use of nuclear power and nuclear science applications. The publications include information covering all policy, technological and management aspects of the definition and implementation of activities involving the peaceful use of nuclear technology. While the guidance provided in IAEA Nuclear Energy Series publications does not constitute Member States' consensus, it has undergone internal peer review and been made available to Member States for comment prior to publication.

The IAEA safety standards establish fundamental principles, requirements and recommendations to ensure nuclear safety and serve as a global reference for protecting people and the environment from harmful effects of ionizing radiation.

When IAEA Nuclear Energy Series publications address safety, it is ensured that the IAEA safety standards are referred to as the current boundary conditions for the application of nuclear technology.

The development of new nuclear power plant designs spans a wide range of alternatives. Some represent minor extensions of current designs, while others incorporate more significant modifications. Terms used to describe designs in various phases of their design and development prompted the IAEA to publish Terms for Describing New, Advanced Nuclear Power Plants (IAEA-TECDOC-936) in 1997. At the time the terms used to describe new designs included advanced designs, next generation designs and evolutionary designs, as well as less technical terms such as passively safe designs, intrinsically safe designs and deterministically safe designs. A precise explanation of the implication of these and similar terms did not exist at the time, and different organizations used the same terms but with different meanings. Such inconsistencies were thought to potentially create confusion, thus resulting in credibility issues. IAEA-TECDOC-936 was aimed at improving the understanding of widely used technical terms in IAEA Member States and providing clarification of their proper usage, as well as similar related terms. In 2005 the IAEA launched the Advanced Reactors Information System on-line database, where terminology used was mainly in accordance with IAEA-TECDOC-936; Safety Related Terms for Advanced Nuclear Plants (IAEA-TECDOC-626); and the IAEA Safety Glossary. Since its publication, the terms in IAEA-TECDOC-936 reflecting the advances of nuclear power plant development in the mid-1990s have been referenced often by IAEA Member States. Technology has advanced since then, and terminology has changed and expanded. Therefore, this publication aims at providing widely used terms for describing advanced nuclear power plants, and clarifying definitions and proper usage. The updates expand on terms by incorporating developments and initiatives since 1997 in the areas of advanced, evolutionary and innovative nuclear reactor designs, including descriptions of design development phases. The terms for describing advanced nuclear power plants of any type should conform to the broad, general understanding by the public as well as by the technical community.

The IAEA staff members responsible for this publication were M. Krause and T. Jevremovic of the Division of Nuclear Power.

CONTENTS

1. INTRODUCTION 1

    1.1. Background 1
    1.2. Objective 2
    1.3. Scope and structure 2
    1.4. Users 2

2. TERMS RELATED TO DESIGN DEVELOPMENT 3

    2.1. Design development phases 3
    2.2. Organizations 4
    2.3. Reactor site and size 4

3. ADVANCED REACTOR DESIGN CATEGORIES 6

    3.1. Advanced design, currently in operation, under construction or licensed 7
    3.2. Evolutionary design 8
    3.3. Innovative design 8
    3.4. Passive design 8
    3.5. Proliferation related terms 9
    3.6. Time related terms 9
    3.7. Technical terms 9
    3.8. Non-technical terms in common use 10

4. REACTOR TYPES 11

    4.1. Water cooled reactors 11
    4.2. Gas cooled reactors 12
    4.3. Molten salt reactors 12
    4.4. Fast reactors 13
    4.5. Small and medium sized or modular reactors 13
    4.6. Microreactors 13
    4.7. Accelerator driven systems 14

5. REACTOR DESIGN PURPOSE 14

    5.1. Commercial 14
    5.2. First of a kind and nth of a kind 14
    5.3. Prototype 14
    5.4. Demonstration 14
    5.5. Experimental 14

6. PERFORMANCE PARAMETERS 15

    6.1. Technical performance 15
    6.2. Economic performance 15

REFERENCES   17
ABBREVIATIONS.   18
CONTRIBUTORS TO DRAFTING AND REVIEW   19
STRUCTURE OF THE IAEA NUCLEAR ENERGY SERIES   20

1. INTRODUCTION

1.1. BACKGROUND

In view of the importance of communication to both the public and to the technical community in general and among the designers of different advanced nuclear reactor lines within the nuclear industry itself, consistency and international consensus are desirable with regard to the terms used to describe various categories of advanced designs. In 1997, the IAEA-TECDOC-936 on Terms for Describing New, Advanced Nuclear Plants [1] was issued and has been widely used since. In 2005, the IAEA launched an online database on Advanced Reactors Information Systems (ARIS) [2] where terminology used was mainly in accordance with that TECDOC and IAEA-TECDOC-626 on Safety Related Terms for Advanced Nuclear Plants [3].

This publication is a revision of IAEA-TECDOC-936 incorporating developments and initiatives since 1997 in the areas of advanced, evolutionary and innovative nuclear reactor designs, description of design development phases, some safety and regulatory terminology, consistent with and complementary to, the newest editions of the IAEA Safety Glossary [4], IAEA Safety Standards Series No. SSR-2/1 (Rev. 1), Safety of Nuclear Power Plants: Design [5], and the IAEA's Radioactive Waste Management Glossary [6]. The terms for describing advanced nuclear power plants (NPPs) need to conform to the broad, general, common understanding by the public as well as by the technical community. Therefore, the terms explained in this publication refer primarily to the state of development of the designs and to the general level of effort needed to bring them to realization, while safety related terms are not the focus of this publication. In this publication the term reactor is synonymous with nuclear reactor, and plant with NPP.

Many organizations have made and continue to develop designs for reactors and systems for improving and advancing nuclear technology. There is a very large spread in the degree of innovation in proposed design approaches and in the corresponding degree of technical maturity of the solutions being achieved or proposed. Although there is also a spread in design objectives ranging from improving performance, economics and safety over what has already been achieved with current technology to expanding the field for application of nuclear energy, strong common threads include enhancement of safety, feedback of experience from operating plants, and incorporation of recent advancements in electronics, computers and human factors. The terms described in this publication are used to distinguish between designs at different phases in development.

The designs considered are NPP designs rather than reactor designs, since the reactor is only a part of the complete installation needed to produce economic, dependable and safe nuclear electricity and other outputs (e.g. hydrogen, heat). Since the previous publication in 1997, the development of advanced NPP designs spanned a wide range of alternatives; some represent small extensions of the present designs while others include significant departures or represent significantly different or new concepts. The terms provided in the IAEA-TECDOC-936 are therefore updated here to incorporate these new developments in NPP designs.

Many of the terms described in this publication have been widely used in some countries, sometimes without sufficiently clear understanding of what they mean and what they imply. Some of these terms have the potential of being misleading to non-experts and of conveying to the public undesirable implications not intended by the designers of advanced NPPs. The criterion for inclusion of each term in the definitions of this publication has been whether the term is already in common, widespread use, not whether such use is desirable. Some terms described here are not compatible with this criterion. They are therefore undesirable, and their use is discouraged; when this is the case the reasons are indicated in their description. Descriptions of some potentially useful terms not now widely used were intentionally omitted to avoid coining or promoting new terms, which, again, would increase rather than reduce potential misunderstanding.

The process of resolving differences resulting from historically different development goals, approaches and timescales in different countries, between varied interests and between differing cultural understanding of words can be difficult; some compromise on an international level was required.

1.2. OBJECTIVE

The objective of this publication is to provide Member States with up-to-date terms for describing advanced NPPs, to draw distinctions between design phases reflecting the maturities of designs, and to clarify definitions of commonly used terms in describing advanced NPPs.

Descriptions here generally conform to definitions but include some elaboration, refinement, and specificity needed to make them applicable and useful for describing advanced NPPs. The overall purpose of these terms is to help:

— Promote the proper use of the terms by members of the nuclear community, rendering the terms more consistent and thereby improving meaningful communication within the technical community and with the public;
— Clarify these terms and thereby achieve a better understanding of the time, effort and investment needed to bring various advanced designs into operation.

An important criterion is clarity, as well as eliminating ambiguity and facilitating ease of application. Anyone who understands the concept of a design should be able to determine quickly and easily whether a design related term conforms to a description. This is more readily achieved by drawing distinctions based on qualitative principles and approaches rather than on quantitative criteria.

Guidance and recommendations provided here in relation to identified good practices represent expert opinion but are not made on the basis of a consensus of all Member States.

1.3. SCOPE AND STRUCTURE

Important terms for describing advanced NPPs are defined, briefly described and put into context in this publication for the following topical areas:

— Design development phases (Section 2);
— Advanced NPP design categories (Section 3);
— Reactor types (Section 4);
— Design purpose (Section 5);
— NPP performance (Section 6).

1.4. USERS

Users of this publication are technical and non-technical people involved in a nuclear power industry.

2. TERMS RELATED TO DESIGN DEVELOPMENT

2.1. DESIGN DEVELOPMENT PHASES

A common understanding of the terms referring to typical phases of NPP design development from conception until completion are described in this section. Although in all cases the work to be done is similar as dictated by the technical requirements, the practice in different countries varies widely in the way the work is divided into phases and the terms used to describe them. Such a breakdown is also strongly influenced by how research and development (R&D), testing and licensing are sequenced into the project.

The design and licensing status are important indicators of the engineering status of an NPP design; that is, its readiness for deployment. An indiscriminate use of various terms to describe design status and differing licensing milestones can lead to confusion and can prevent a clear understanding of the real status. To approach this problem, a classification model based on practices in some IAEA Member States is shown in Fig. 1.

In this design classification model, the NPP design status is assessed against a set of often used milestones with these four broad typical phases:

— Concept description;
— Conceptual design;
— Basic design;
— Detailed design, either for bounding site conditions (IVa) or site specific (IVb).

Figure 1 shows only typical engineering activities in each of these four phases. In view of the different practices in different countries, a consensus on this terminology and scope may be difficult to achieve. Achieving a consensus on corresponding terminology for R&D, testing and licensing is also difficult, and therefore, no attempts are made to include those aspects.

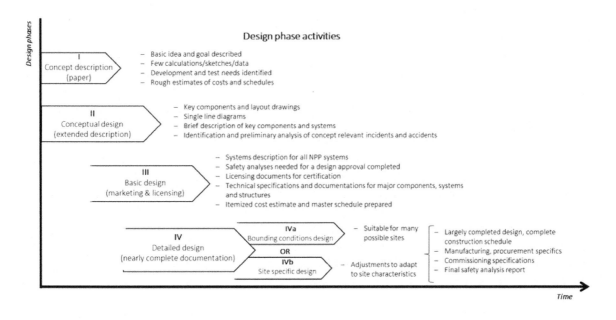

FIG. 1. Phases and activities during design development (excluding major testing).

2.2. ORGANIZATIONS

Organizations that are in various ways involved with NPP designs use terminology related to the design. The most relevant organizations are, in order of their involvement in the design process:

Technology holder / vendor / design organization: owns the intellectual property for the design of an NPP and may employ an engineering team to further develop and refine the design. Depending on the extent of design progress, they may be expected to provide a suitable design envelope to the development company such that licences and permits can be satisfactorily obtained.

Development company: finalizes the site specific design, procures and constructs the NPP. May obtain the licence to operate the NPP. May become, or subsequently form part of, the operating organization.

Supply chain: delivers services throughout the design phases, pre-construction, manufacture, construction and startup phases or supply materials, manufactured components, fuel supply and fuel cycle services.

Operator / operating organization: "Any *person or organization* applying for *authorization* or authorized and/or responsible for *safety* when undertaking *activities* or in relation to any *nuclear facilities* or *sources* of *ionizing radiation*" [4].

Owner: an organization that is established to own the NPP. Owner and operator roles are often performed by a single organization that is responsible for securing the finance for the NPP and is responsible for NPP safety.

Utility: typically, the electric power company that receives or buys electricity from an NPP, or an organization that receives another product such as desalinated water, heat, hydrogen or medical isotopes. Often it is the same company as the NPP owner or operator.

Licence holder / licensee: holds the licence to construct/operate an NPP.

Regulator / regulatory body: "An authority or a system of authorities designated by the government of a State as having legal authority for conducting the regulatory *process*, including issuing *authorizations*, and thereby regulating the *nuclear, radiation, radioactive waste* and *transport safety*" [4]. "The holder of a current licence is termed a licensee. A licence is a product of the *authorization* process, although the term licensing process is sometimes used" [6].

Nuclear energy programme implementing organization (NEPIO): a mechanism, which may involve high level and working level committees, to coordinate the work of government, owner, operator and regulator in the nuclear power infrastructure development [7].

2.3. REACTOR SITE AND SIZE

Important terminology related to the location, major components and size of advanced NPPs, in alphabetical order:

Balance of plant (BOP): NPP remaining structures, systems and components that comprise a complete NPP and are not included in the nuclear power conversion system or, in the case of water cooled reactors (WCRs), the nuclear steam supply system.

Containment building / containment: the containment structure and the systems with the functions of isolation, control and management of mass and energy releases, control and limitation of radioactive releases, and control and management of combustible gases. Details of the containment structure, systems and components can be found in Ref. [1].

Emergency planning zone (EPZ): emergency planning for the protection of NPP personnel, emergency workers and the public beyond the site boundary is a necessary element of overall NPP safety and provides an additional level of defence in depth [8]. The required EPZ depends on local regulations; therefore, a design may claim to have a small or zero EPZ, but this may not be realizable in all applications. According to Ref. [4], the EPZ consists of two parts: the precautionary action zone and the urgent protective action planning zone:

> **"precautionary action zone (PAZ).** An area around a *facility* for which *emergency arrangements* have been made to take *urgent protective actions* in the *event* of a *nuclear or radiological emergency* to avoid or to minimize *severe deterministic effects* off the site. *Protective actions* within this area are to be taken before or shortly after a *release* of *radioactive material* or an *exposure*, on the basis of prevailing conditions at the *facility.*

> **"urgent protective action planning zone (UPZ).** An area around a *facility* for which *arrangements* have been made to take *urgent protective actions* in the event of a *nuclear or radiological emergency* to avert *doses* off the site in accordance with international *safety standards. Protective actions* within this area are to be taken on the basis of *environmental monitoring* — or, as appropriate, prevailing conditions at the *facility.*"

Microreactors: very small reactors within the small modular reactor category[1].

Modular construction: a method where reactor systems or subsystems are prefabricated for onsite assembly, to various degrees of modularization, all aiming to accelerate the overall construction schedule.

Modular design: in relation to SMRs, an NPP that consists of one or more essentially identical units (modules) that are fully or partly factory prefabricated for onsite installation or assembly, or, in relation to large reactors, prefabricated modules of the nuclear island and BOP systems.

Nuclear power plant/ plant: a facility that includes one or more reactors to convert nuclear energy into usable power; also an electrical or thermal generating facility that uses a nuclear reactor as its heat source.

Nuclear reactor / reactor: an engineered system, other than a nuclear weapon, designed or used to sustain nuclear fission in a self-supporting chain reaction [9]. Although there are many types of nuclear reactors, they all incorporate certain essential features including the use of fissionable material as fuel, a moderator (such as water) to increase the likelihood of fission (unless reactor operation relies on fast neutrons), a reflector to conserve escaping neutrons, coolant provisions for heat removal, instruments for monitoring and controlling reactor operation, and protective devices (such as control rods and shielding).

Nuclear steam supply system (NSSS) / nuclear island / reactor module: the combination of reactor core, reactor coolant system, and related auxiliary systems including the emergency core cooling system, decay heat removal system and chemical volume and control system [9]. In the context of some SMR or microreactor designs, a reactor module is a complete nuclear island unit fabricated in a factory and transported to site for installation.

Power reactor: a reactor designed to produce electrical or thermal energy.

Reference unit / reactor: the closest existing (operating or under construction) unit of the same design type on which the advanced design is based. Regulators may have their own specific meaning of this term.

Site / site area / site footprint: "A geographical area that contains an *authorized facility, authorized activity* or *source,* and within which the management of the *authorized facility* or *authorized activity* or first responders may directly initiate *emergency response actions*" [4]. This is typically the area within the security perimeter fence or other designated property marker, such as the site boundary fence. It is also a basis on which NPP fees are calculated [9] and equivalent to the land use for energy generation.

Small advanced reactors: reactors within the SMR category which produce electricity of typically up to about 300 MWe per unit or module [2].

Unit (single-, dual-, multi-unit): each unit represents a separate reactor (nuclear island and BOP) capable of being operated. In the case of dual- or multi-unit plants, a unit can operate independently of the

---

[1] Small modular reactors are a subset of small and medium sized or modular reactors (SMRs).

state of completion or operating condition of any other units co-located on the same site but in different containment/confinement buildings, even though the units may have some shared or common systems.

3. ADVANCED REACTOR DESIGN CATEGORIES

Nuclear power plant designs that are being developed span a wide range of alternatives; some represent very small extensions of current designs, others incorporate more significant modifications, and still others depart very markedly from current designs [2].

The relationship between the design category related terms is shown in Fig. 2. Some operating and (at time of this publication) proposed NPP designs are categorized as advanced NPP designs (Sections 3.1 and 3.2) if they are of interest and/or with merit. These can be further characterized, depending upon the need for additional development (engineering, R&D, demonstration NPP). The degree of innovation of designs increases from small in the engineering-only category to unlimited in the innovative designs (Section 3.3).

The full spectrum of advanced NPP designs or concepts for which current interest or merit can be identified covers current designs, evolutionary designs as well as designs requiring substantial development efforts, such as innovative designs. Fig. 3 sketches the relative development efforts and cost for advanced designs vs. departure from current designs. The designs in both evolutionary and innovative categories need engineering and may also need R&D and confirmatory testing prior to finalizing the design of the first evolutionary reactor or of the prototype and/or demonstration reactor for innovative designs. The amount of such R&D and confirmatory testing depends on the degree of both the innovation to be introduced and the related work already done, or the experience that can be built upon.

Evolutionary designs involve only moderate modifications or improvements over current designs with a strong emphasis on maintaining proven designs. In this way, commercial risks are minimized. If modifications and design changes are larger, with more departure from current designs, and with introduction of unproven features, risks increase correspondingly since less or no operating experience exists. Many SMR designs aim to reduce this risk as much as possible, given the design evolutions or innovations involved.

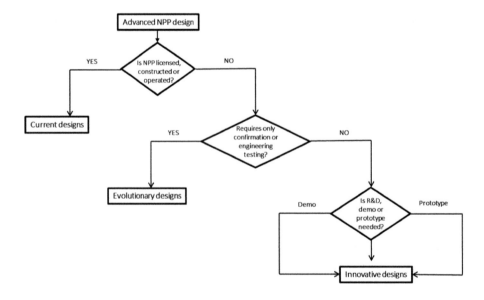

FIG. 2. Relationship between terms related to NPP design categories (all terms used in this figure are described in Section 3).

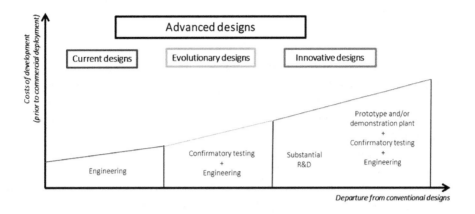

FIG. 3. Development effort and cost for advanced designs vs. departure from current designs.

All NPPs in operation not belonging to the main advanced NPP designs category are identified as commercial power reactors, i.e. those deployed as *n*th of a kind until the 1980s.

3.1. ADVANCED DESIGN, CURRENTLY IN OPERATION, UNDER CONSTRUCTION OR LICENSED

Advanced design: a design of current interest for which substantial improvement over its predecessors and/or commercial power reactor designs have been realized. Advanced designs, currently in operation, under construction or licensed [10] include WCRs that are improved from commercial power reactor WCRs. They differ from evolutionary designs in that their detailed design is completed (Figs 2 and 3).

Advanced designs that have already been built and operated or are under construction [10], belong to the advanced NPP design category and are referred to in this publication as current designs. These are mainly improved commercial power reactor WCR designs, which still comprise the majority of the operating NPPs. They were developed within advanced reactor design development programmes in various countries through the 1980s and 1990s, reflecting advanced safety features based on lessons learned from the Three Mile Island and Chernobyl accidents, new users' requirements derived during the last several decades of operational experience and, in some cases, newly emerged licensing requirements. Among them, the advanced boiling water reactor (BWR), an advancement of the conventional BWR, was developed by Japan and the United States of America (USA), and built and operated in Japan. The AP-1000, an advanced conventional pressurized water reactor (PWR), had been developed in the USA, built and operated in China and is being built in the USA. Another example is the European pressurized water reactor (EPR), an advancement of the conventional PWR, which had been developed in Europe and has been built and operated in China and is under construction in France, Finland and UK. The APR-1400, an advancement of the conventional PWR that had been developed in the Republic of Korea, is built and operated there and in the United Arab Emirates. The VVER[2]-1200, an advanced design of the VVER-1000, is under construction in the Russian Federation, Bangladesh and Turkey. The BN800, an evolution of the BN600, is an example of a non-water-cooled advanced fast reactor design in the Russian Federation. Designs identified as current designs under the main advanced NPPs designs category are all the reactors described in the recent IAEA booklet on Advanced Large Water Cooled Reactors: A Supplement to: IAEA Advanced Reactors Information System (ARIS) 2020 Edition [11].

---

[2] VVER: water cooled, water moderated power reactor

3.2. EVOLUTIONARY DESIGN

Evolutionary design: an advanced design that achieves improvements over current designs through small to moderate modifications, with a strong emphasis on maintaining proven design elements to minimize technological risks. They are near-term deployable reactors of water cooled, gas cooled or liquid metal cooled technology, typically under certification or licensing for which there is a lack of experience and for which, as a consequence, safety regulation may need to be revised or developed, and consequently there would be a need to reach a consensus on these needs at the international level (i.e. IAEA safety standards). The development of an evolutionary design requires, at most, engineering and confirmatory testing prior to deployment (Figs 2 and 3).

When considering future designs, it is also necessary to ask whether any candidate design is of current interest or merit. Among the many possible designs, there are some which have been previously developed (either wholly or partially) and then abandoned. Many of these have been conceived and considered but not found to be of sufficient interest for further development, and presumably some remain to be conceived and evaluated. In the approach taken for this publication, none of these can currently be considered advanced NPP designs; that designation applies only to designs of current interest or merit which, upon completion of their development, are expected to incorporate improvements of varying degrees and kinds over current NPPs.

Design efforts on evolutionary designs aim to achieve improvements over current designs through small to moderate modifications, often guided by a utility requirements document (URD) with the goal to improve safety, reduce costs and licensing uncertainties. Some designs achieve this through simplification, larger safety margins, improved severe accident prevention and mitigation, longer grace periods [4] in emergency situations, improvement of the human–machine interface systems, shorter construction time and improved maintainability.

3.3. INNOVATIVE DESIGN

Innovative design: an advanced design which incorporates conceptual changes in design approaches or system configuration in comparison with existing practice [2]. Substantial R&D, feasibility tests, and possibly a prototype or demonstration reactor are required prior to deployment (Figs 2 and 3).

The range of designs for which substantial development efforts are still needed is much wider than for the category of evolutionary designs. For some concepts, development is almost completed, while for other concepts much work remains to be performed.

The key attribute of an innovative design is that it is based on conceptual changes in design approaches or system configuration in comparison with established practice.

3.4. PASSIVE DESIGN

The IAEA publication TECDOC-626 [3] remains the most common reference when classifying passive components or systems; it suggests four categories. The following definitions are provided based on Refs [1] and [4]:

Passive component: "A *component* whose functioning does not depend on an external input such as actuation, mechanical movement or supply of power" [4].

Passive system: an automatically or manually initiated safety system that is provided to ensure that the required safety functions are achieved. Either a system which is composed entirely of passive components and structures, or a system which uses active components in a very limited way to initiate subsequent passive operation [1].

Passive design: a reactor that relies entirely on passive systems for accident prevention and mitigation.

Passive safety feature(s): this term should be used only in conjunction with specific examples of passive systems or components. In general, a safety feature (for design extension conditions) is defined as an *"Item that is designed to perform a safety function for or that has a safety function for design extension conditions"* [4]. This term should be used only in conjunction with specific examples of passive systems or components.

3.5. PROLIFERATION RELATED TERMS

Proliferation resistant design: though not an official IAEA safeguards term, in common use refers to deploying nuclear energy systems in a way to reduce the risk of nuclear weapons proliferation. The basic principle for proliferation resistance requires that intrinsic factors (that result from the technical design) and extrinsic measures (state commitments, obligations and policies) of proliferation resistance be implemented throughout the full life cycle of the system to ensure that the system will be an unattractive means of acquiring fissile material or technology for a nuclear weapons programme [12].

Safeguards by design: the process of including international safeguards considerations throughout all phases of an NPP's life cycle, from the initial conceptual design to construction and into operations including design modifications and decommissioning [13].

3.6. TIME RELATED TERMS

Although the terms future design and next generation design are straightforward and not usually subject to misuse or misunderstanding, they are qualitative and should only be used with an appropriate indication regarding the time frame, if it is not derived from the context.

Future design: can refer to new concepts or those that represent significant technological advancements for improved safety, economics and resource utilization. Since this term can cover a wide spectrum, it is suggested that it be used with great care and only with a definition specific to the context.

Next generation design: this term implies time or specific characteristics, or both. Applied to NPP designs, it can cover a wide spectrum ranging from modest modifications over their predecessors, to concepts with radical and fundamental changes that are far more ambitious than those for evolutionary designs. On the other hand, the term is often used by industry for a new series of NPPs which may be only a decade apart as opposed to the much longer time usually needed for radically new concepts. Similarly, a national nuclear power programme may refer to deployment of a next generation design as part of its forward programme. The term may be used to reinforce the point that state of the art technology is being adopted, this being different from technology currently deployed within the programme or elsewhere. Usage of the term 'next generation design' to describe both extremes is acceptable as long as the respective meaning is clear from the context.

Near term deployment: this often used term is only meaningful if used in relation to the Member State's nuclear power programme plans. The IAEA's Nuclear Reactor Technology Assessment for Near Term Deployment [14] describes near term deployment within the IAEA's Milestones in the Development of a National Infrastructure for Nuclear Power [7] framework, meaning advanced designs that are expected to be deployable in time for construction according to the Member State's timeline.

3.7. TECHNICAL TERMS

Proven design: a design that has been proven in equivalent applications or is to a large extent based on an operating NPP.

Integral and integrated design: refers primarily to a PWR design in which all major components of the reactor primary circuit — including pressurizer, steam generators/heat exchangers, and in some cases coolant pumps and control rod drive mechanism — are enclosed in a single reactor vessel. For molten salt reactors (MSR) or fast reactors (FR) the meaning is similar but includes different components.

Semi-integral design: also called compact loop design, this refers to a reactor system design in which major components in the primary circuit such as pressurizers, coolant pumps, and steam generators/heat exchangers are directly flanged to the reactor pressure vessel without piping. Although some of those components may be placed inside of the reactor vessel in a semi-integral design, if all the major components are placed inside of the reactor vessel, it is called integral design.

Direct cycle design: an NPP design where the primary coolant (steam or a gas) heated in the reactor is directly transferred to the turbine. Steam generators or heat exchangers are not required in these designs.

Loop type design: an NPP design which has primary system components (i.e. steam generator, primary coolant pump and pressurizer) connected to each other and the reactor pressure vessel through large pipes. This design may consist of different numbers of loops — typically two, three or four.

Pool type design: the design in which the reactor core is immersed in a pool of coolant (water or liquid metal).

Below ground design: an innovative nuclear design in which the entire nuclear island is located below ground for safety, security and/or economic reasons.

Floating design: an NPP design located on a floating platform at sea. There are largely two types of floating NPPs: one is on a fixed platform, which is still floating but is connected to seabed like offshore oil platforms and is expected to stay in one place for a long period of time; the other is placed on a mobile platform such as a barge or a ship, which can move by itself or by being towed and is expected to change its location as needed according to its application (e.g. electricity, heat).

3.8. NON-TECHNICAL TERMS IN COMMON USE

Certain non-technical terms have been used to describe concepts for which it is claimed that all accident sequences that could potentially lead to unacceptable consequences have been practically eliminated by design. The use of these terms for a specific design could be taken as a confirmation that other designs are insufficient in that regard.

The following terms (in alphabetical order) have been used but are discouraged in technical documentation without being substantiated, because they do not have an accepted general or technical meaning:

Catastrophe free design / deterministically safe design: a term sometimes used for concepts in which all accident sequences, even those of very low probability and leading to unacceptable consequences, are practically eliminated by design measures. The use of this descriptor for an entire NPP or its reactor is discouraged since it implies absolute safety, which is unrealistic. Also, it may cause confusion since the use of deterministic methods constitutes an important element of established licensing practice. In fact, all operating NPPs can be said to be deterministically safe within their licensing basis. Finally, the use of this term for a specific design could be taken as an indication that other designs are far from being catastrophe free. Hence, this descriptor should not be used.

Forgiving design / fault tolerant design: sometimes used instead of *passive design*, which is described in Section 3.4. The use of 'forgiving' as NPP descriptor should be avoided. Fault tolerance is typically associated with instrumentation and control systems only [4] and should also be avoided to describe a design in its entirety.

Inherently safe design: in accordance with [3], the unqualified use of this term should be avoided for describing an entire NPP or its reactor.

Revolutionary design: sometimes used to characterize an advanced design that is substantially different from evolutionary design. It has essentially the same attributes as an innovative design, but since the word 'revolutionary' may have a negative connotation, the use of this term should be avoided.

Walk away safe design: NPPs that rely entirely on passive systems for an indefinite time for accident prevention and mitigation, not requiring any kind of operator action. The potential use of this term could imply that operators would walk away in case of an accident. Hence the 'walk away' descriptor should not be used.

4. REACTOR TYPES

As shown in Fig. 4, a multitude of combinations of fuel, moderator and coolant materials underpin the names of commonly used reactor types. Advanced and innovative reactor designs exist in almost every one of these generic reactor types.

Existing NPPs and development programmes cover light water reactor (LWR), heavy water reactor (HWR), high temperature gas cooled reactor (HTGR), and liquid metal reactor (LMR) technologies; in addition, MSR designs have been developed and some are in advanced development phases. Some of these technologies have been further developed at smaller scales and are collectively called SMRs. Brief definitions of nuclear reactor types are given below.

4.1. WATER COOLED REACTORS

Water cooled reactors have been the cornerstone of the nuclear industry in the twentieth century and advanced WCRs continue to play a central role well into the twenty-first century.

4.1.1. Pressurized water reactors/water-water energetic reactors

Pressurized water reactors or VVER produce steam for the turbine in steam generators that are connected to the reactor pressure vessel by large pipes, called hot-legs and cold-legs.

4.1.2. Boiling water reactors

Boiling water reactors use the steam produced inside the core directly in the steam turbine.

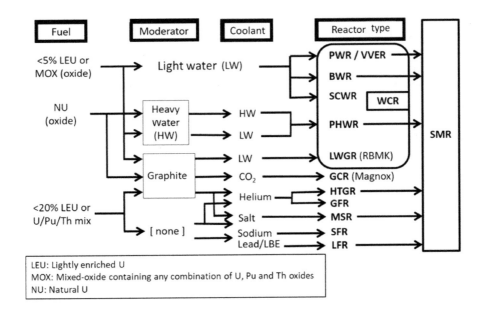

FIG. 4. Nuclear reactor type by fuel, moderator, and coolant.

4.1.3. Pressurized heavy water reactors/heavy water reactors

Pressurized heavy water reactors (PHWR) or HWRs use enriched water, the molecules of which comprise hydrogen atoms that are made up to more than 99% of deuterium, a heavier hydrogen isotope. This heavy water, used as a moderator, improves the overall neutron economy, allowing fuel to be used that does not require enrichment. Some HWR designs, however, do use slightly enriched uranium fuel for improved economics or better resource utilization.

4.1.4. Supercritical water reactors

To improve NPP thermal efficiencies and economics, R&D for supercritical water reactors (SCWRs) is being pursued. Supercritical water exists at temperatures and pressures above its critical point, where the liquid and vapour states are indistinguishable. This water is commonly used in advanced coal, oil and gas fired power plants. Plant efficiencies of SCWRs are expected to be around 1.3 times higher than in WCRs.

4.2. GAS COOLED REACTORS

Gas cooled reactors currently represent about 3% of the total number of reactors in commercial operation worldwide, most of which are advanced carbon dioxide gas cooled reactors (AGRs) in the UK that will be phased out in the next decades.

4.2.1. High temperature gas cooled reactors

High temperature gas cooled reactors use a thermal spectrum reactors and graphite as the moderator and reflector, are helium cooled, and make use of coated particle fuel. The basic concept is not new and prototype HTGRs have been licensed, built and operated; the last two prototype NPPs, Fort St. Vrain (USA) and the THTR-300 (Germany), were both shut down in the 1980s.

4.2.2. Modular high temperature gas cooled reactor

Modular high temperature gas cooled reactors (MHTGRs) follow the philosophy from the 1980s in Germany (development of the high temperature reactor module) and the USA to develop a passively safe HTGR. To achieve the safety objectives, the design relies on the inherent high temperature characteristics of tristructural isotropic (TRISO) coated fuel particles along with passive heat removal capability from the large height-to-diameter ratio, low power density core in an uninsulated steel reactor vessel. No active circulation or even the need for the coolant is credited in the safety case. The ultimate heat sink is via a core cavity cooling system (a cooler outside, around the reactor vessel with passive features) or in most cases the building structure (concrete) and surrounding earth. China completed HTR-PM is the first example of an MHTGR.

4.3. MOLTEN SALT REACTORS

Molten salt reactors have molten salt(s) as either the reactor coolant or both the fuel and the coolant. Most MSRs have a thermal neutron spectrum, while molten salt fast reactors (MSFRs) have a fast spectrum.

4.4. FAST REACTORS

An FR is a reactor in which little or no neutron moderator is used, and the fission chain reaction is primarily sustained by fast neutrons. The fast neutron spectrum in an FR can largely increase the energy yield from uranium resource as compared to thermal reactors.

4.4.1. Liquid metal cooled fast reactor

A liquid metal cooled fast reactor (LMFR) is an FR that uses liquid metal as coolant. Examples of the liquid metal used for these designs as the primary coolant are sodium (in a sodium fast reactor, SFR) and lead and lead-bismuth eutectic (in a lead cooled fast reactor, LFR). Sodium cooled technology is mature and has been used for >50 years, while lead and lead-bismuth eutectic coolants are proposed for some innovative designs.

4.4.2. Molten salt fast reactor

A molten salt fast reactor is an FR in which the reactor coolant or both the fuel and the reactor coolant are molten salt(s).

4.4.3. Gas cooled fast reactor

A gas cooled fast reactor (GFR) is an FR which uses helium as a reactor coolant with the objective of maximizing fuel utilization through high thermal efficiency.

4.5. SMALL AND MEDIUM SIZED OR MODULAR REACTORS

Small and medium sized or modular reactors are an option to fulfil the need for flexible power generation for a wider range of users, grid sizes and applications. For small and medium sized reactors, small sized means up to 300 MWe, while medium sized means newer generation reactors with power between 300 MWe and 700 MWe. A small modular reactor is defined as an advanced reactor that produces electricity of up to 300 MWe per module and is designed either as a single or multimodule NPP (Section 2.3), which systems and components can be fabricated as modules in their factory setting then transported to site to shorten construction duration. The US Nuclear Regulatory Commission defines SMRs for the purposes of calculating fees as the class of LWRs having a licensed thermal power rating less than or equal to 1 000 MWt per module. This rating is based on the thermal power equivalent of a light water SMR with an electrical power generating capacity of 300 MWe or less per module [15].

4.6. MICROREACTORS

Microreactor is a relatively new term for a reactor able to produce between 1 and 30 MW of thermal energy used directly as heat or converted to electric power. Components are factory fabricated and assembled and transported to the location of deployment. Microreactors self-adjust in all operational conditions and use passive safety systems to prevent overheating or core melt. They are expected to operate for years without refuelling, but many designs require fuel with a higher enrichment than 5%.

4.7. ACCELERATOR DRIVEN SYSTEMS

Accelerator driven system (ADS) is an innovative concept of a hybrid system for the transmutation of long lived radioisotopes. The ADS consists of a high power proton accelerator, a heavy metal spallation target that produces neutrons when bombarded by the high power beam, and a subcritical reactor core that is neutronically coupled to the spallation target.

5. REACTOR DESIGN PURPOSE

5.1. COMMERCIAL

Commercial reactors are reactors in power plants that have been built to full scale and intended solely for commercial use in the generation of electricity and/or process heat for industrial applications or other non-electric products [2].

5.2. FIRST OF A KIND AND NTH OF A KIND

FOAK is an acronym for 'first of a kind,' in the sense of first of many, not a once-off prototype or demonstration. It is used for the first reactor put in operation using a new technology or design. It can cost significantly more than later units called NOAK, an acronym for 'nth of a kind,' which have incorporated lessons learned from the earlier ones.

5.3. PROTOTYPE

A prototype reactor is the first physical reactor from which future commercial reactors are developed. It may be at a reduced scale or lacking some systems (e.g. turbine generator) and is intended to demonstrate overall reactor performance, reliability, safety, and economics.

5.4. DEMONSTRATION

A demonstration reactor is the practical exhibition of how the advanced design performs. This may be a partial or complete reactor design either at reduced or full scale but is intended to demonstrate the effective and safe operation of the design's evolutionary or innovative features.

5.5. EXPERIMENTAL

Typically, the first design of a new innovative reactor technology built for the purpose of validating the performance of reactor core materials and fuels, exploring safety limits and uncertainties, and gaining critical lessons so that the technology may be licensed and commercialized at some future time.

6. PERFORMANCE PARAMETERS

6.1. TECHNICAL PERFORMANCE

Breeding ratio (BR): ratio of final fissile material produced in a fast or thermal reactor to the initial loaded fissile content. Breeding gain is the excess fuel produced. If the BR<1, the reactor is a burner — meaning it consumes more fissile material than it produces; if BR>1, the reactor is a breeder; if BR=1 or the breeding gain is zero, this reactor is called an iso-breeder — meaning it produces the same amount of fuel as it consumes during operation.

Burnup: integrated energy produced from the depletion of fissile material, including fissile material produced, while the fuel resides in the reactor core, divided by the fresh fuel total mass of uranium (MWd/kgU or GWd/tU).

Capacity: amount of electric power (MWe) or other non-electric product that an NPP or a unit can produce.

Capacity factor: ideal energy supply capability of an NPP divided by the energy output that would be produced if it operated at its rated power output for a typical year, or over the entire lifetime operation. Generally, it is expressed as percentage. In the IAEA PRIS database [10] the term 'load factor' is used, which includes production losses due to demand changes.

Design life: "time during which a *facility* or *component* is expected to perform according to the technical specifications to which it was produced" [4]. For an NPP this is the real time in years of NPP operation while all structures, systems and components remain qualified to perform their functions. This may include major refurbishment or replacement of systems or components during shutdowns, depending on the NPP's design philosophy. Another term sometimes used is 'effective full power years' to describe the lifetime of in-core components that deteriorate from irradiation.

Efficiency: ratio of product output over the thermal power of the reactor. For electricity, this is expressed as MWe(net) / MWth(core).

Enrichment: wt% of ^{235}U in the fuel. Enriched uranium contains a greater mass percentage of ^{235}U than the naturally occurring uranium of 0.72% [6].

Grace period: "time during which a *safety function* is ensured in an *event* with no necessity for action by personnel" [4].

Load following / load cycling: NPP operation during which the electrical output is adjusted throughout the day to match the demand and/or for grid frequency control. Such power manoeuvring capability, typically expressed as a power range and rate of power change, is beneficial in a grid with large portion of intermittent generators.

6.2. ECONOMIC PERFORMANCE

Expenditures for NPP projects start a long time before the start of construction and include costs related to engineering, R&D, testing, concept safety demonstration and licensing.

Terms often used to describe the economics of an NPP project are defined as follows, in order of relevance to the overall lifetime cost:

Levelized cost of electricity (LCOE): total cost to build, operate and decommission an NPP over its lifetime divided by the total electricity output dispatched from the NPP (e.g. US $/kWh). This value depends on the following factors, approximately in the order as they are listed below:

Capital cost, total construction costs: include the cost of site preparation, construction, manufacturing, commissioning, financing (interest during construction), cost escalation and contingency costs of an NPP.

Overnight construction costs: base construction cost plus applicable owners' costs, contingency and first core costs. It is referred to as an 'overnight' cost in the sense that time value costs are not included. Costs included are those before revenues are accrued and are exclusive of finance charges accruing during the construction period.

Construction time: most of the time this is the time from the beginning of construction (i.e. pouring of the first nuclear island concrete) up to NPP entries into commercial operations, but other events are also used (i.e. to first criticality, to first grid connection, etc.).

Cost of capital: finance charges for the invested capital, expressed as a percentage.

Interest during construction: finance charges accruing during the construction period.

Discount rate: rate of interest that is used to calculate the present value of an amount of money that is expected to be received or paid in the future. It should reflect the risk of the project to which it is applied.

Factory construction: manufacture of specific equipment or complete modules for the nuclear energy system in a facility equipped with suitable tooling. Offers a higher level of efficiency, repeatability, process control, quality assurance and standard of work environment compared to site based construction.

Operation and maintenance cost: cost of operating an NPP, including its maintenance (and in some cases fuel cost and charges for decommissioning and waste management).

Fuel cost: ongoing cost of fuel during operations, often expressed as a percentage of operations cost (and in some cases includes charges for spent fuel).

Fuel cycle costs: costs in addition to fuel costs, related to any reprocessing or advanced fuels. While these are insignificant for existing LWRs, they may be significant for innovative designs, such as an MSR or an FR.

First core costs: the cost of the first full fuel load to enable startup of a reactor.

REFERENCES

[1]   INTERNATIONAL ATOMIC ENERGY AGENCY, Terms for Describing New, Advanced Nuclear Power Plants, IAEA-TECDOC-936, IAEA, Vienna (1997).

[2]   INTERNATIONAL ATOMIC ENERGY AGENCY, ARIS Database, IAEA, Vienna, https://aris.iaea.org

[3]   INTERNATIONAL ATOMIC ENERGY AGENCY, Safety Related Terms for Advanced Nuclear Plants, IAEA-TECDOC-626, IAEA, Vienna (1991).

[4]   INTERNATIONAL ATOMIC ENERGY AGENCY, IAEA Nuclear Safety and Security Glossary: Terminology Used in Nuclear Safety, Nuclear Security, Radiation Protection and Emergency Preparedness and Response, 2022 (Interim) Edition, IAEA, Vienna (2022).

[5]   INTERNATIONAL ATOMIC ENERGY AGENCY, Safety of Nuclear Power Plants: Design, IAEA Safety Standards Series No. SSR-2/1 (Rev. 1), IAEA, Vienna (2016).

[6]   INTERNATIONAL ATOMIC ENERGY AGENCY, Radioactive Waste Management Glossary, 2003 Edition, IAEA, Vienna (2003).

[7]   INTERNATIONAL ATOMIC ENERGY AGENCY, Milestones in the Development of a National Infrastructure for Nuclear Power, IAEA Nuclear Energy Series No. NG-G-3.1 (Rev. 1), IAEA, Vienna (2015).

[8]   INTERNATIONAL ATOMIC ENERGY AGENCY, Managing Siting Activities for Nuclear Power Plants, IAEA Nuclear Energy Series No. NG-T-3.7 Rev.1, IAEA, Vienna (2022).

[9]   INTERNATIONAL ATOMIC ENERGY AGENCY, Design of the Reactor Containment and Associated Systems for Nuclear Power Plants, IAEA Safety Standards Series No. SSG-53, IAEA, Vienna (2019).

[10]   INTERNATIONAL ATOMIC ENERGY AGENCY, PRIS Database, IAEA, Vienna, https://pris.iaea.org/PRIS/home.aspx

[11]   INTERNATIONAL ATOMIC ENERGY AGENCY, Advanced Large Water Cooled Reactors: A Supplement to: IAEA Advanced Reactors Information System (ARIS) 2020 Edition, IAEA, Vienna (2020), https://aris.iaea.org/Publications/20-02619E_ALWCR_ARIS_Booklet_WEB.pdf

[12]   INTERNATIONAL ATOMIC ENERGY AGENCY, Options to Enhance Proliferation Resistance of Innovative Small and Medium Sized Reactors, IAEA Nuclear Energy Series No. NP-T-1.11, IAEA, Vienna (2014).

[13]   INTERNATIONAL ATOMIC ENERGY AGENCY, International Safeguards in the Design of Nuclear Reactors, IAEA Nuclear Energy Series No. NP-T-2.9, IAEA, Vienna (2014).

[14]   INTERNATIONAL ATOMIC ENERGY AGENCY, Nuclear Reactor Technology Assessment for Near Term Deployment, IAEA Nuclear Energy Series No. NR-T-1.10 (Rev. 1), IAEA, Vienna (2022).

[15]   NUCLEAR REGULATORY COMMISSION, NRC 10 CFR § 170.3 Definitions, US Nuclear Regulatory Commission, Washington, DC, https://www.nrc.gov/reading-rm/doc-collections/cfr/part170/part170-0003.html

ABBREVIATIONS

ADS	accelerator driven system
AGR	advanced carbon dioxide gas cooled reactor
ARIS	Advanced Reactors Information System
BOP	balance of plant
BWR	boiling water reactor
EPR	European pressurized reactor
EPZ	emergency planning zone
FOAK	first of a kind
FR	fast reactor
GFR	gas cooled fast reactor
HTGR	high temperature gas cooled reactor
HWR	heavy water reactor
LCOE	levelized cost of electricity
LFR	lead cooled fast reactor
LMFR	liquid metal cooled fast reactor
LMR	liquid metal reactor
LWR	light water reactor
MHTGR	modular high temperature gas cooled reactor
MSFR	molten salt fast reactor
MSR	molten salt reactor
MWe	megawatt electricity
NEPIO	nuclear energy programme implementing organization
NOAK	nth of a kind
NPP	nuclear power plant
NSSS	nuclear steam supply system
PAZ	precautionary action zone
PHWR	pressurized heavy water reactor
PWR	pressurized water reactor
R&D	research and development
SCWR	supercritical water reactor
SMR	small and medium sized or modular reactor
TRISO	tristructural isotropic
UPZ	urgent protective action planning zone
URD	utility requirements document
VVER	water cooled, water moderated power reactor
WCR	water cooled reactor

CONTRIBUTORS TO DRAFTING AND REVIEW

Banoori, S.	Pakistan Atomic Energy Commission, Pakistan
Bilic Zabric, T.	International Atomic Energy Agency
Delmastro, D.	National Atomic Energy Commission, Argentina
Ganda, F.	International Atomic Energy Agency
Jevremovic, T.	International Atomic Energy Agency
Karseka-Yanev, T.	International Atomic Energy Agency
Krause, M.	International Atomic Energy Agency
Kriventsev, V.	International Atomic Energy Agency
Mathers, D.	Nuclear Innovation Research Office, United Kingdom
Memmott, M.	Brigham Young University, United States of America
Moriwaki, M.	Hitachi-GE Nuclear Energy, Ltd., Japan
Reitsma, F.	International Atomic Energy Agency
Subki, M.H.	International Atomic Energy Agency
Whitlock, J.	International Atomic Energy Agency

Consultants Meeting
Vienna, Austria: 29–30 July 2020

Structure of the IAEA Nuclear Energy Series*

Nuclear Energy Basic Principles
NE-BP

Nuclear Energy General Objectives
NG-O

1. Management Systems
NG-G-1.#
NG-T-1.#

2. Human Resources
NG-G-2.#
NG-T-2.#

3. Nuclear Infrastructure and Planning
NG-G-3.#
NG-T-3.#

4. Economics and Energy System Analysis
NG-G-4.#
NG-T-4.#

5. Stakeholder Involvement
NG-G-5.#
NG-T-5.#

6. Knowledge Management
NG-G-6.#
NG-T-6.#

Nuclear Reactor** Objectives
NR-O

1. Technology Development
NR-G-1.#
NR-T-1.#

2. Design, Construction and Commissioning of Nuclear Power Plants
NR-G-2.#
NR-T-2.#

3. Operation of Nuclear Power Plants
NR-G-3.#
NR-T-3.#

4. Non Electrical Applications
NR-G-4.#
NR-T-4.#

5. Research Reactors
NR-G-5.#
NR-T-5.#

Nuclear Fuel Cycle Objectives
NF-O

1. Exploration and Production of Raw Materials for Nuclear Energy
NF-G-1.#
NF-T-1.#

2. Fuel Engineering and Performance
NF-G-2.#
NF-T-2.#

3. Spent Fuel Management
NF-G-3.#
NF-T-3.#

4. Fuel Cycle Options
NF-G-4.#
NF-T-4.#

5. Nuclear Fuel Cycle Facilities
NF-G-5.#
NF-T-5.#

Radioactive Waste Management and Decommissioning Objectives
NW-O

1. Radioactive Waste Management
NW-G-1.#
NW-T-1.#

2. Decommissioning of Nuclear Facilities
NW-G-2.#
NW-T-2.#

3. Environmental Remediation
NW-G-3.#
NW-T-3.#

(*) as of 1 January 2020
(**) Formerly 'Nuclear Power' (NP)

Key
BP: Basic Principles
O: Objectives
G: Guides and Methodologies
T: Technical Reports
Nos 1–6: Topic designations
#: Guide or Report number

Examples
NG-G-3.1: Nuclear Energy General (**NG**), Guides and Methodologies (**G**), Nuclear Infrastructure and Planning (topic 3), **#1**
NR-T-5.4: Nuclear Reactors (**NR**), Technical Report (**T**), Research Reactors (topic **5**), **#4**
NF-T-3.6: Nuclear Fuel (**NF**), Technical Report (**T**), Spent Fuel Management (topic 3), **#6**
NW-G-1.1: Radioactive Waste Management and Decommissioning (**NW**), Guides and Methodologies (**G**), Radioactive Waste Management (topic 1) **#1**

20

 **IAEA**
International Atomic Energy Agency

ORDERING LOCALLY

IAEA priced publications may be purchased from the sources listed below or from major local booksellers.

Orders for unpriced publications should be made directly to the IAEA. The contact details are given at the end of this list.

NORTH AMERICA

Bernan / Rowman & Littlefield

15250 NBN Way, Blue Ridge Summit, PA 17214, USA

Telephone: +1 800 462 6420 • Fax: +1 800 338 4550

Email: orders@rowman.com • Web site: www.rowman.com/bernan

REST OF WORLD

Please contact your preferred local supplier, or our lead distributor:

Eurospan Group

Gray's Inn House
127 Clerkenwell Road
London EC1R 5DB
United Kingdom

Trade orders and enquiries:

Telephone: +44 (0)176 760 4972 • Fax: +44 (0)176 760 1640
Email: eurospan@turpin-distribution.com

Individual orders:

www.eurospanbookstore.com/iaea

For further information:

Telephone: +44 (0)207 240 0856 • Fax: +44 (0)207 379 0609
Email: info@eurospangroup.com • Web site: www.eurospangroup.com

Orders for both priced and unpriced publications may be addressed directly to:

Marketing and Sales Unit
International Atomic Energy Agency
Vienna International Centre, PO Box 100, 1400 Vienna, Austria
Telephone: +43 1 2600 22529 or 22530 • Fax: +43 1 26007 22529
Email: sales.publications@iaea.org • Web site: www.iaea.org/publications